SCIENTIFIC AMERICAN EDUCATIONAL PUBLISHING

SCIENTIFIC AMERICAN INVESTIGATES LIFE CYCLES

THE STRANGE LIFE CYCLE OF A MAYFLY

ANNA MCDOUGAL

Published in 2025 by The Rosen Publishing Group
in association with Scientific American Educational Publishing
2544 Clinton Street, Buffalo NY 14224

Portions of this work were originally authored by Greg Roza and published as *The Bizarre Life Cycle of a Mayfly.*
All new material in this edition authored by Anna McDougal.

Library of Congress Cataloging-in-Publication Data

Names: McDougal, Anna, author.
Title: The strange life cycle of a mayfly / Anna McDougal.
Description: Buffalo, NY : Scientific American Educational Publishing, [2025] | Series: Scientific American investigates life cycles | Includes bibliographical references and index.
Identifiers: LCCN 2024010933 (print) | LCCN 2024010934 (ebook) | ISBN 9781725350915 (library binding) | ISBN 9781725350908 (paperback) | ISBN 9781725350922 (ebook)
Subjects: LCSH: Mayflies–Juvenile literature. | Mayflies–Life cycles–Juvenile literature.
Classification: LCC QL505 .M33 2025 (print) | LCC QL505 (ebook) | DDC 595.7/34156–dc23/eng/20240318
LC record available at https://lccn.loc.gov/2024010933
LC ebook record available at https://lccn.loc.gov/2024010934

Designer: Tanya Dellaccio Keeney
Editor: Caitie McAneney

Photo credits: Cover Erik Karits/Shutterstock.com; p. 5 Sandra Standbridge/Shutterstock.com; p. 7 Melinda Fawver/Shutterstock.com; p. 9 Rostislav Stefanek/Shutterstock.com; p. 11 Achkin/Shutterstock.com; p. 13 Mladen Mitrinovic/Shutterstock.com; p. 15 Tommy Studio/Shutterstock.com; p. 17 Jaywchk/Shutterstock.com; p. 10 Diane Isabel/Shutterstock.com; p. 21 Kirsanov Valeriy Vladimirovich/Shutterstock.com.

Printed in the United States of America

CPSIA compliance information: Batch #CSSA25. For Further Information contact Rosen Publishing at 1-800-237-9932.

CONTENTS

Words in the glossary appear in **bold** type the first time they are used in the text.

SHORTLIVED

Mayflies are truly strange. One day, they're here! They're rising out of the water and **swarming** in the air. They're stuck to the sides of buildings and trying to get in your windows. The next day, they're gone.

The life of an adult mayfly is very short. The scientific order mayflies belong to is called Ephemeroptera (ih-feh-muh-RAHP-tuh-ruh). This word comes from the Greek words for "short-lived" and "wings." Follow the life cycle of this incredible insect!

Mayflies can be very beautiful up close!

FUN FACT

SOME MAYFLY SWARMS ARE SO LARGE THEY SHOW UP ON WEATHER RADAR!

MAYFLY FEATURES

You may be able to **identify** a mayfly in your own backyard. Adult mayflies are easy to spot. They have one pair of large, triangle-shaped wings and one pair of smaller wings. They have a long body and six long legs. Most mayflies have three caudal filaments, or tails, at the end of their long body. These features make them great flyers.

There are more than 3,000 species, or kinds, of mayflies. North America alone is home to more than 600 species of mayfly!

If you see a mayfly sitting on a tree, you usually find hundreds more—often on the same tree!

FUN FACT

MAYFLIES LIVE IN AND NEAR WATER ON EVERY CONTINENT EXCEPT ANTARCTICA.

LARVAE UNDERWATER

Mayflies need water. That's because they lay their eggs there. After mayfly eggs **hatch** in water, the **larvae** are called nymphs. Nymphs have gills along the sides of their body. Gills allow the nymphs to breathe underwater.

Shortly after hatching, the nymphs often find something to hang on to. Different species act differently. Some tunnel into sand or mud. Some crawl between rocks. A few even swim! They stay in the water until adulthood, which takes up to a year.

Nymphs don't have wings like adult mayflies.

FUN FACT

MAYFLY LARVAE ARE OFTEN FOUND IN FAST-MOVING STREAMS, BUT THEY MAY ALSO LIVE IN LAKES AND PONDS.

GROWING UP

Depending on the species, nymphs can grow to between 0.16 and 1.2 inches (4 and 30 mm) long. As they grow, the nymphs **molt** many times. The nymph increases in size with each molt. The number of molts is different for each species, but it ranges from 20 to 50 over the course of a year.

As nymphs grow, they become male or female. Males have very long front legs that they use when **mating**. A young mayfly's body becomes more like an adult's with each molt.

This mayfly nymph is molting, leaving behind its skin.

FUN FACT

MOST MAYFLY NYMPHS EAT THE REMAINS OF PLANTS AND ANIMALS. SOME EAT LIVE PLANTS AND ALGAE.

SUPER SUBIMAGOS

What happens when a nymph molts for the last time? It gets ready to leave the water. For a nymph's final molt, some of its insides are emptied out of its body. Air replaces the guts, which are no longer needed since the insect won't eat again.

When a nymph reaches the surface, its body cracks open and its wings come out! The insect floats on the water until it's strong enough to fly. Now the mayfly is a subimago!

Subimago is a mayfly's first stage of adulthood.

FUN FACT

NYMPHS OFTEN LET GO OF THEIR UNDERWATER HOMES AT ABOUT THE SAME TIME AND FLOAT TO THE WATER'S SURFACE.

AMAZING IMAGOS

When subimagos are able to fly, they leave the water. But that's not the end of their story. They swarm onto land to find new resting spots. These may be on a tree or long grass or under a bridge. Here they prepare to molt again.

Unlike most insects, mayflies pass through two adult stages during their life cycle. The mayfly's final molt allows its legs and body to grow a little longer. After molting, the subimagos become imagos. Mayflies are the only insects that molt after their wings have formed!

This is a male mayfly.

FUN FACT

IMAGOS ARE ABLE TO MATE. IT'S THE LAST STAGE OF THEIR LIFE CYCLE.

A SUDDEN SWARM

Once mayflies become imagos, you are likely to see them swarm. In areas where mayflies are common, people often find thick clouds of them.

Swarms of female imagos mix in with the male swarms. Then, each female spreads hundreds of eggs over water. Some release them directly into water. Soon after, the adult mayflies die. The time between the moment a subimago **emerges** from the water and the time the imago dies is less than two days. Often it's much shorter.

This building is overrun with mayflies!

FUN FACT

MAYFLIES ARE OFTEN DRAWN TO LIGHTS, SUCH AS STREETLIGHTS.

PART OF THE FOOD CHAIN

Mayflies have many **predators** in each part of their life cycle. Mayfly eggs are eaten by snails and other insect larvae. Nymphs are **prey** for fish, frogs, birds, insects, and roundworms. Adults are food for fish, birds, and bigger insects.

Because of this, each female mayfly can lay up to about 1,200 eggs. This is important for mayfly **survival**. Once a subimago leaves the water, it has far fewer predators to worry about.

Fly fishermen use **bait** that looks like a mayfly. That's because some fish love to eat mayflies!

FUN FACT

ADULT MAYFLIES DON'T EAT. HOWEVER, THEY'RE FOOD FOR MANY OTHER ANIMALS.

IS IT A PEST?

Do mayflies help or hurt? Many people think they are pests. The swarms may seem disgusting to those who don't like insects. They can even make it hard to see while driving! Their bodies often pile up after they die.

However, mayflies are an important part of the **environment**. They are food for many animals. They are a sign that the water in a place is healthy. And they don't bite people. Next time you see a mayfly swarm, marvel at their short, strange life!

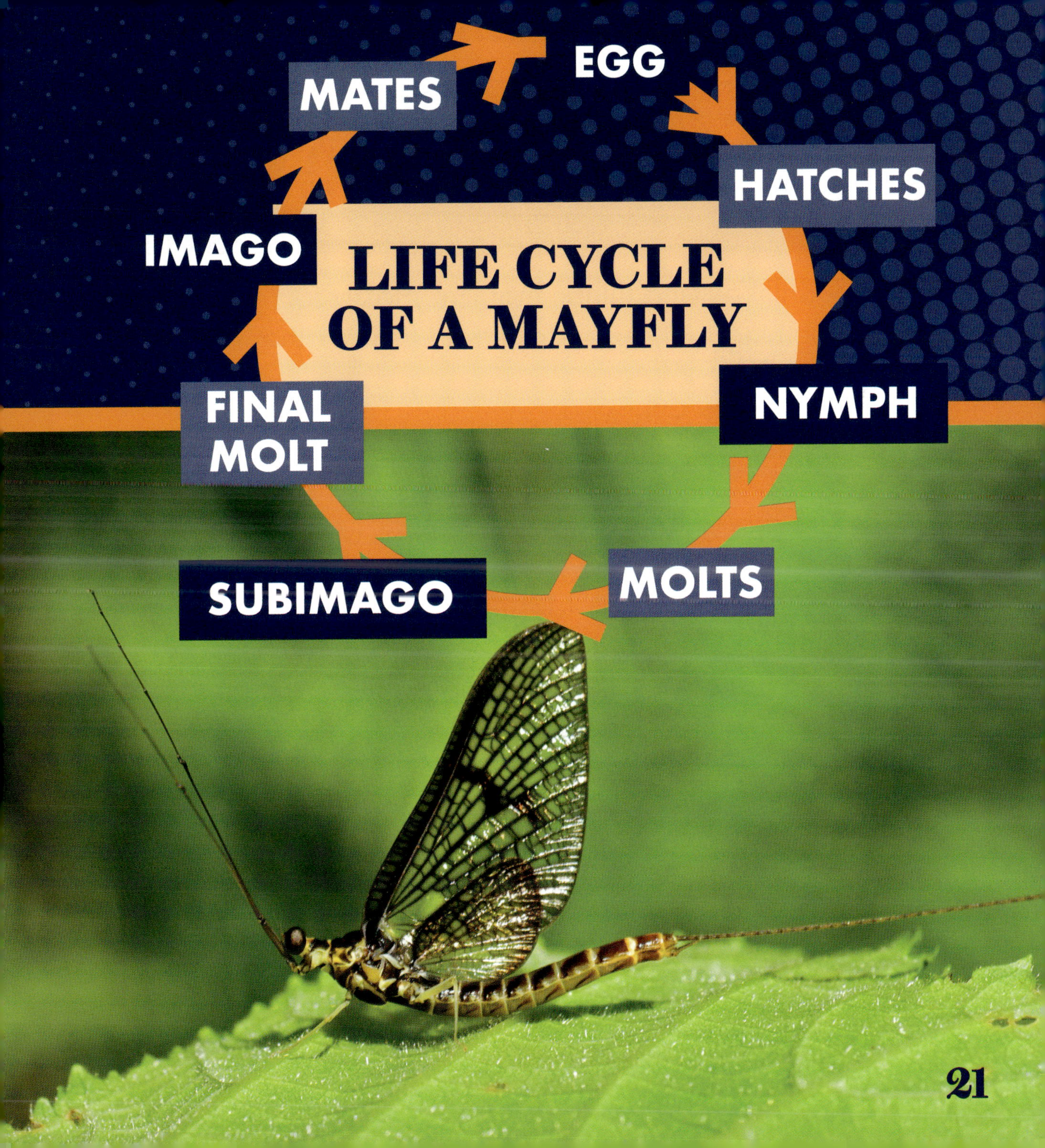
LIFE CYCLE OF A MAYFLY
EGG
HATCHES
NYMPH
MOLTS
SUBIMAGO
FINAL MOLT
IMAGO
MATES

GLOSSARY

algae: Plantlike living things that are mostly found in water.

bait: Something that's used to attract fish or other animals so they can be caught.

emerge: To come out.

environment: The natural world.

hatch: To break out of.

identify: To find out the name or features of something.

larvae: Bugs in an early life stage that often have a wormlike form. The singular form is "larva."

mate: To come together to make babies.

molt: To shed old shells, skin, fur, or feathers while a new covering grows in.

predator: An animal that hunts other animals for food.

prey: An animal that is hunted by other animals for food.

radar: A way of using radio waves to find distant objects.

survival: The ability to stay alive.

swarm: A large group of flying insects. Also, to fly in a large group.

FOR MORE INFORMATION

Books

Ganeri, Anita. *Insects and Invertebrates.* London, UK: Wayland Publishing, 2021.

Grey, Susan Heinrichs. *Mayfly.* Ann Arbor, MI: Cherry Lake Publishing, 2021.

Websites

Mayflies

www.nwf.org/Educational-Resources/Wildlife-Guide/Invertebrates/Mayflies
Explore facts about mayflies with the National Wildlife Federation.

Mayfly

kids.britannica.com/students/article/mayfly/275750
Learn more about the amazing mayfly!

Publisher's note to educators and parents: Our editors have carefully reviewed these websites to ensure that they are suitable for students. Many websites change frequently, however, and we cannot guarantee that a site's future contents will continue to meet our high standards of quality and educational value. Be advised that students should be closely supervised whenever they access the internet.

INDEX